BEE BALM GARDENING HORTICULTURIST GUIDE FROM CULTIVATION TILL COMMMERCIAL SUCCESS

Bee Balm Mastery, Ultimate Guide To Cultivating, Marketing, And Profiting From Your Garden's Herbal Treasure Trove

MANUEL SHELTON

Copyright © 2024 by Manuel Shelton

All rights reserved. No part of this publication may be reproduced, distributed, or transmitted in any form or by any means, including photocopying, recording, or other electronic or mechanical methods, without the prior written permission of the publisher, except in the case of brief quotations embodied in critical reviews and certain other noncommercial uses permitted by copyright law.

Disclaimer

The views expressed in this book are solely those of the author and do not necessarily reflect the views of any company, organization, or individual.

 The author is not engaged in any endorsement deals or partnerships with any entities mentioned in this book. Any references to products, services, or organizations are for informational purposes only and do not constitute endorsement.

Readers are encouraged to conduct their own research and exercise their own judgment before making any decisions based on the information provided in this book

Contents

CHAPTER ONE 15
SETTLEMENTS FOR BEE BALM GARDENING 15
- Choosing The Ideal Site For Bee Balm 15
- Tips For Preparing Soil.. 17
- Recognizing Your Needs For Water And Sunlight 19
- Keeping An Eye On Soil Moisture 20

CHAPTER TWO 23
ESTABLISHING BEE BALM 23
- Selecting The Ideal Planting Time 23
- Step-By-Step Guidelines For Planting 24
- Advice On Bee Balm Transplanting 26

CHAPTER THREE 29
CARE AND MAINTENANCE OF BEE BALM 29
- Guidelines For Fertilisation And Watering 29
- Pruning Methods For Vigorous Development 30
- Typical Insect And Disease Problems And Solutions 31

CHAPTER FOUR 35
PROPAGATION METHODS 35
- Techniques For Propagating Seeds 35

   Separation And Root Clippings 37
   FAQ And Troubleshooting For Propagation 39
CHAPTER FIVE 41
  CULLING AND APPLYING BEE BALM 41
   Knowing When To Gather 41
   Methods For Gathering Bee Balm 42
   Medicinal And Culinary Applications Of Bee Balm 43
CHAPTER SIX 47
  INCLUDING BEES IN THE DESIGN OF LANDSCAPE 47
   Selecting The Ideal Site: 47
   Design Suggestions For Using Bee Balm 49
   Ideas For Companion Planting 53
   Bee Balm: A Tool For Bee-Friendly Garden Design
   56
CHAPTER SEVEN 61
  IN COMMERCIAL GARDENING, BEE BALM 61
   Growing Up: Methods Of Commercial Cultivation 61
   Promoting Products Made With Bee Balm 64
   Regulatory And Legal Aspects 67
CHAPTER EIGHT 71
  COMMONLY ASKED QUESTIONS OR FAQS 71

Common Questions Concerning Bee Balm 73

Troubleshooting Manual For Typical Problems 75

Professional Hints And Counsel 78

CHAPTER NINE 81

COMMON QUESTIONS AND ASSISTANCE 81

Overcoming Typical Obstacles In Bee Balm Gardening 81

Techniques For Achievement 83

Resources For Additional Education And Assistance 86

CONCERNING THIS BOOK

"Bee Balm Gardening" is an invaluable resource for horticulturists of all experience levels as well as eager novices, providing a thorough look into the fascinating world of Monarda, or bee balm. Readers are exposed to the essence of bee balm gardening from its earliest beginnings in this finely crafted text, which also explores the variety, relevance, and critical role of this plant in improving garden landscapes.

The voyage commences with an engrossing Overview of Bee Balm, revealing the essence of this extraordinary plant. The essentials are explained to readers, including its botanical identity, the significant role it plays in gardening ecosystems, and a perceptive synopsis of its many variants. Now that the reader has this basic understanding, they are ready to set off on their Bee Balm Gardening adventure.

"Getting Started with Bee Balm Gardening" is a must-have resource for anybody interested in gardening. It provides crucial information on location selection, sophisticated soil preparation methods, and an understanding of the complex relationship between water and sunshine requirements. Equipped with this understanding, readers may proceed with assurance when planting bee balm, with careful guidance on time, methods, and transplantation advice.

"Bee Balm Care and Maintenance" becomes apparent as the adventure goes on, providing light on the vital procedures of watering, fertilizing, and trimming, which are necessary for raising healthy bee balm specimens. With this knowledge, readers may protect their flower partners from frequent pests and diseases and prolong their lives.

The investigation delves deeper into the field of propagation methods, equipping readers with the skills necessary to divide, propagate seeds, and take root cuttings to preserve the beauty of bee balm. Helpful help in the form of troubleshooting tips and an extensive FAQ section turns unexpected setbacks into growth opportunities.

"Harnessing the Harvest: Culinary and Medicinal Uses of Bee Balm" reveals the many uses for this miracle of nature, from its highly valued medicinal qualities to its enticing culinary uses. The story smoothly moves into the field of landscape design, including advice, suggestions for companion planting, and methods for making vibrant, bee-friendly gardens.

"Bee Balm in Commercial Gardening" is a guide for business owners in the field of commercial gardening, explaining the subtleties of expanding growing methods, successful marketing approaches, and negotiating legal and regulatory environments.

A testament to its comprehensiveness, "Bee Balm Gardening" ends with a lengthy FAQ section that deftly answers frequently asked questions, solves problems, and offers a wealth of resources for additional education and assistance. With grace, wisdom, and constant excitement, this book illuminates the route toward effective bee balm production, making it a valuable resource for both novice and seasoned gardeners.

Overview of Bee Balm

Bee balm, a stunning and adaptable flowering plant in the mint family, is also known by its scientific name, Monarda. Bee balm enhances the visual appeal and perfume of any garden setting with its vivid flowers and fragrant foliage.

Native to North America, this perennial herbaceous plant is adored by gardeners for its many advantages and simple upkeep.

Bee Balm: What Is It?

Bright tubular flowers that draw pollinators like bees, butterflies, and hummingbirds are the distinguishing features of bee balm. There are many different tints of these flowers, such as red, pink, purple, and white, so gardeners have a lot of possibilities to select from. Bee balm is prized for its therapeutic qualities in addition to its visual beauty. The plant's leaves are a popular herbal cure for a variety of illnesses because they may be made into a calming tea with antibacterial and antifungal qualities.

Bee Balm's Significance in Gardening

Bee balm is highly valued in gardening for several reasons, chief among them being its capacity to draw helpful pollinators into the garden. Pollination of flowering plants is vital for the development of fruit and seeds, and bees and other pollinating insects are an important part of this process. Bee balm is a plant that you may use to attract pollinators to your garden, which will benefit the ecosystem's general health and biodiversity.

Moreover, bee balm is a low-maintenance plant that grows well in a variety of environments. It works well in a variety of garden settings since it can take both full sun and partial shade and is adaptable to different types of soil. It's also a great option for novice gardeners or those who want hassle-free plants for their landscape because of its resistance to pests and diseases.

A Summary of Bee Balm Types

Gardeners can choose from a variety of bee balm cultivars, each with their special qualities and advantages. Several well-liked variations include:

Balm for Scarlet Bees, Monarda didyma

The most well-known type of balm is probably scarlet bee balm, which is prized for its vivid red blooms that liven up any garden. Typically reaching a height of two to four feet, this cultivar attracts a multitude of pollinators with its nectar-rich blossoms that bloom from mid to late summer.

Fistulosa Monarda (Wild Bergamot)

Another well-liked bee balm species with lavender-pink blooms and fragrant foliage is wild bergamot. It grows naturally in meadows and plains throughout North America. This cultivar is prized for its culinary and medicinal properties in addition to its aesthetic appeal.

Citronella monarda (lemon bee balm)

As its name implies, lemon bee balm is highly valued for its zesty aroma and delicate lavender blossoms. Because of its lovely scent, this species is frequently used in herbal teas and potpourris. Due to its diminutive stature—lemon bee balm usually reaches a height of 1-2 feet—it is a good choice for container gardening and tiny gardens.

Spotted Bee Balm (Monarda punctata)

The unusual yellow-spotted bracts around its pale pink flowers are what setspotted bee balm apart from other plants. Native to eastern North America, this type is frequently found growing in rocky or sandy soils. Gardeners love spotted bee balm for its distinctive blossoms and capacity to draw pollinators.

Adding a range of bee balm species to your garden will increase biodiversity, help local pollinator populations, and create visual interest. You're sure to enjoy the beauty and advantages that these adaptable plants have to offer, whether you decide to plant lemon bee balm for its delightful smell or scarlet bee balm for its eye-catching red blossoms.

CHAPTER ONE

SETTLEMENTS FOR BEE BALM GARDENING

Monarda, or bee balm, is a pleasant addition to any garden because of its propensity to draw pollinators such as hummingbirds, butterflies, and bees, in addition to its bright blossoms.

To make sure your plants flourish, it's imperative to grasp the fundamentals before delving into the world of bee balm gardening. Here is a thorough how-to to get you going:

Choosing The Ideal Site For Bee Balm

Selecting the ideal location for your bee balm plants is essential to their general health and growth. The following are some important things to remember:

Sunlight Requirements: Bee balm is a versatile choice for a variety of garden settings since it thrives in full sun to moderate shade. Ideally, for best growth and flowering, try to get at least six hours of sunshine each day. Offering some midday shade will help keep the plants from wilting if you reside in a really hot region.

Soil Quality: Well-drained soil with a pH between that of a slight acid and neutral is preferred by bee balm. Avoiding soggy or compacted soil is crucial since these conditions can cause root rot and other problems. Enhance the texture and fertility of the soil by adding organic matter, such as compost, before planting.

Air movement: To avoid common fungal illnesses like powdery mildew, adequate air movement is essential. Make sure there is enough distance between plants when choosing a place so that adequate ventilation may occur. Bee balm should not be planted in densely populated or extremely protected regions, particularly if your area has a severe humidity problem.

Closeness to Other Plants: Although bee balm is usually innocuous, it's important to take into account how well it gets along with nearby plants. Plant it apart from aggressive spreaders and plants that have different needs for sunlight and moisture. Rather, go with companion plants like salvia, echinacea, or lavender, which grow in similar conditions and enhance the aesthetics of bee balm.

Tips For Preparing Soil

One of the most important steps in laying the groundwork for a successful bee balm planting is preparing the soil. To give your seedlings the greatest start possible, heed these tips:

Cleaning the Area:

To start, remove any weeds, stones, or other objects that could obstruct root growth or compete with the plants for nutrients from the planting area. To create a loose, friable texture and break up any compacted areas, loosen the soil with a shovel or garden fork.

Bringing the Soil Back:

To increase the structure and fertility of your soil, you might need to amend it with organic matter, depending on the makeup of your soil. The top few inches of soil can benefit from the addition of compost, aged manure, or peat moss to improve drainage, hold on to moisture, and supply vital nutrients for plant development.

Checking the pH:

To establish the ideal growing conditions for bee balm, think about testing the pH of the soil to see if any changes are required. To guarantee nutrient availability and root health, aim for a pH range of 6.0 to 7.0, which is slightly acidic to neutral. Lime can be added to raise the pH if the soil is too acidic, and sulfur can be used to lower the pH if the soil is too alkaline.

Mulching: To assist retain moisture, controlling weed growth, and controlling soil temperature, spread an organic mulch layer around the base of the bee balm plants after the soil has been prepared and planted. Mulching further enriches the growing environment by gradually adding organic matter to the soil as it decomposes.

Recognizing Your Needs For Water And Sunlight

For bee balm plants to grow vigorously and bloom profusely, it is imperative to maintain a balance between the plants' requirements for water and sunlight. What you should know is as follows:

Light: Bee balm grows best in full sun to light shade, while the precise amount of light it needs varies based on soil moisture content and environment. For best development and flowering, try to get at least six hours of direct sunlight each day. In warmer climates, midday shade helps shield plants from intense heat and lessen water stress.

Watering: Bee balm needs regular watering, but if the soil is wet for an extended length of time, it might lead to root rot. To help newly planted bee balm grow their root systems, water them frequently enough to keep the soil evenly moist but not soggy. Bee balm is comparatively drought-tolerant once established; protracted dry spells may be the only times it needs extra watering.

Keeping An Eye On Soil Moisture

Regular monitoring of soil moisture levels is crucial to avoid overwatering or underwatering. Examine the soil at the base of the plants with your finger; if it feels dry, it's time to water.

When at all possible, avoid watering from above as this raises the danger of foliar infections. Instead, use a soaker hose or drip irrigation system to irrigate the base of the plants, which will allow water to reach the roots directly.

Mulching: By lowering evaporation and weed competition, adding a layer of organic mulch around the base of bee balm plants can help preserve soil moisture.

Mulching also aids in controlling soil temperature, protecting roots from temperature swings in colder climates and keeping them cool in warmer weather. Throughout the growth season, add more mulch as needed to the layer of mulch made of shredded bark, straw, or compost.

CHAPTER TWO

ESTABLISHING BEE BALM

Selecting The Ideal Planting Time

The optimal time to plant bee balm is essential to its healthy development and growth. In general, early spring or late autumn are the ideal seasons to plant bee balm.

The weather is milder throughout these seasons, which helps plants establish their roots without being overly stressed by heat.

Planting can also be done in late winter in areas with mild weather. However, as these harsh conditions might impair the plant's ability to thrive, it is imperative to avoid planting bee balm during the height of summer or in cold temperatures.

Step-By-Step Guidelines For Planting

Choose a Good Site: Pick a spot that gets full to partial sunshine for your bee balm plants. Since bee balm prefers well-drained soil, make sure the planting location has adequate drainage.

Before planting, prepare the soil by pulling weeds and tilling the ground down to a depth of approximately 6 to 8 inches. For better soil texture and fertility, add organic matter to the soil, such as old manure or compost.

Planting Bee Balm Seeds: If growing bee balm from seeds, put them straight into the prepared soil, making sure to space them according to the seed packet's recommended spacing. Gently pat down the seeds after lightly covering them with a thin coating of dirt.

When transplanting bee balm seedlings, make sure the holes you dig in the prepared soil are just a little bit bigger than the seedlings' root balls. Make sure the seedlings are planted at the same depth as they were in their original containers, place them into the holes, and cover them with soil.

Watering: To assist the soil surrounding the roots settle after planting, give the bee balm a good soak. During the early stages of growth, keep the soil continuously damp but not soggy.

Mulching: To assist keep moisture in the soil, inhibit weed growth, and control soil temperature, apply a layer of mulch around the base of the plants.

Maintenance: Keep a regular eye out for any indications of pests or illnesses on the bee balm plants.

For optimal development and blooming, give extra water during dry spells and fertilize as needed.

Advice On Bee Balm Transplanting

If done correctly, transplanting bee balm can be a simple task. The following advice can help to guarantee a successful transplant:

Timing: When the weather is cooler and the plants are under less stress, it is best to transplant bee balm in the early spring or late autumn.

Prepare the New Location: Make sure the new planting site has adequate solar exposure and well-drained soil before transferring plants there.

Digging Up the Plant: Be sure to excavate around the root ball of the plant before removing bee balm from its current site, being cautious not to damage the roots.

Carefully Transplanting:

Take caution when handling the plant to prevent breaking its roots. Plant the bee balm at the same depth as it was growing at its new site.

Watering: To aid in the bee balm's establishment in its new site, give it copious amounts of water after transplanting.

Maintain a constant moisture content in the soil in the early days following transplantation.

Give Support: Until the transplanted bee balm becomes established in its new place, give it support if needed to keep it from toppling over.

You can effectively plant and transplant bee balm in your garden and ensure lovely blossoms and healthy growth by following these instructions and advice.

CHAPTER THREE

CARE AND MAINTENANCE OF BEE BALM

Monarda, or bee balm, is a colorful and adaptable flowering plant that can enhance the aesthetics and usefulness of your garden. To guarantee that your bee balm flourishes and keeps drawing pollinators like bees, butterflies, and hummingbirds, proper upkeep and care are essential. We'll cover all the important facets of bee balm upkeep and care in this guide to assist you in creating a productive garden.

Guidelines For Fertilisation And Watering

Your bee balm plants need to be watered often to stay healthy and vibrant, especially in the sweltering summer months. The soil should always be somewhat damp but not soggy. Watering deeply once or twice a week, allowing the water to reach the root zone of the soil, is a good general rule of thumb.

Bee balm should ideally be watered from the base up to prevent overwatering the leaves, which raises the possibility of fungal illnesses.

For delivering water straight to the roots, think about utilizing a drip irrigation system or soaker hose.

Bee balm typically doesn't need much fertilizer in the form of heavy feeding. The nutrients needed for strong development and plenty of flowers can be obtained by applying a balanced, slow-release fertilizer in the early spring as new growth begins to appear. Keep in mind that overfertilizing might result in excessive growth of foliage at the price of blooms.

Pruning Methods For Vigorous Development

Bee balm upkeep requires pruning, which keeps the plant from getting too dense and disease-prone while also encouraging rapid growth. Frequent pruning also promotes the growth of fresh flowers and keeps the garden looking neat.

After the first flush of blossoms has faded, in late spring or early summer, trim spent flower heads and any diseased or dead foliage with clean, sharp pruning scissors.

Deadheading is a technique that enhances a plant's look while refocusing its energy on blooming new blossoms.

Furthermore, it helps to deliberately remove older, woody growth from the base of the plant to thin down overloaded stems. This lowers the chance of fungal diseases by allowing sunlight and air movement to reach the center of the plant.

Typical Insect And Disease Problems And Solutions

Although bee balm is a robust and resilient plant in general, it can occasionally become harmed by pests and diseases that affect its look and overall health. Maintaining the health of your bee balm depends on your ability to see these problems early on and take prompt action to resolve them.

Common pests that can harm bee balm include powdery mildew, spider mites, and aphids. An intense water blast can be used to remove aphids and spider mites from plants, or insecticidal soap can be used to manage their presence. Examine the undersides of leaves frequently for indications of infestation.

The fungus powdery mildew frequently causes a white, powdery coating to develop on bee balm leaves.

Make sure there is enough airflow around the plants by spacing them appropriately and refraining from watering them from above to control powdery mildew. Consider using a fungicidal spray that is intended to control powdery mildew if the illness continues.

You may have a flourishing bee balm garden that not only enhances the beauty of your landscape but also offers vital habitat for pollinators by according to these watering and fertilizing rules, using appropriate pruning procedures, and being on the lookout for pests and diseases.

Your bee balm plants will repay you with colorful blooms year after year if you give them a little love and care.

CHAPTER FOUR

PROPAGATION METHODS

Techniques For Propagating Seeds

Growing bee balm plants from seed is an interesting and rewarding process. Make sure you get premium seeds from a reliable source before you begin. Starting with small pots or a seed tray, fill them with a well-draining potting mix. Before planting the seeds, lightly moisten the soil.

Plant the bee balm seeds, pressing them lightly into the soil's surface. It is important to remember that seeds need light to sprout, so don't bury them too deeply. Make sure the seeds stay moist by misting the soil's surface after sowing.

To create a climate similar to a greenhouse and assist keep moisture, cover the tray or pots with a plastic dome or wrap them in plastic.

The soil may dry out too fast if the seed tray or pots are placed in direct sunlight. Instead, place them in a warm, bright area.

Make sure the soil is constantly damp but not soggy. Depending on the type and environmental factors, germination usually takes seven to twenty-one days.

The seedlings can be moved into separate pots filled with potting mix once they have sprouted multiple true leaves, which normally happens in a few weeks. To prevent breaking the fragile seedlings' roots, handle them gently. Till the seedlings are big enough to be placed outside, keep them growing in a warm, bright area.

Select a sunny spot in the garden with well-draining soil when transferring seedlings there. When the plants are fully grown, they should be spaced 12 to 18 inches apart. To aid in their establishment in their new surroundings, give the recently transplanted seedlings plenty of water.

Separation And Root Clippings

Bee balm plants can be multiplied successfully by division and root cuttings, which work best in the spring or autumn when the plants are not actively growing.

First things first: pick a mature, robust bee balm plant with a strong root system.

Dig up the plant gently, being sure to cover the entire root ball to prevent injuring the roots, and to proliferate by division.

Shake off any extra dirt gently to reveal the roots. Divide the plant into smaller portions using a clean, sharp knife or garden spade, being sure to leave several healthy branches and roots attached to each division.

After dividing, plant the parts again, being sure to put them at the same depth as when they were growing before. To aid in the plants' adjustment to their new locations, give the freshly divided plants plenty of water.

Choose robust, meaty roots from the parent plant for root cuttings. Thick roots are ideal. Cut the roots into parts that are two to three inches long with a sharp knife. Make sure to cover the root cuttings with a thin layer of soil when you plant them horizontally in the prepared soil.

Till new growth appears, water the freshly planted root cuttings and maintain a continually moist soil. It could take a few weeks for root cuttings to take hold, so exercise patience and keep giving them the care they need.

FAQ And Troubleshooting For Propagation

What is the duration required for bee balm seeds to sprout?

A: Depending on the kind and growing environment, bee balm seeds usually sprout in 7 to 21 days. To encourage successful seedling emergence, make sure the soil stays continuously moist during the germination phase.

What's preventing my bee balm seeds from sprouting?

Several variables, such as inadequate soil moisture, an erroneous planting depth, or low-quality seed, might influence the germination of seeds.

Make sure the soil stays continuously damp but not soggy, then plant the seeds at the proper depth according to the instructions on the seed packet.

Additionally, using premium, fresh seeds from a reliable supplier might increase germination rates.

Q: After splitting, my bee balm plants are wilting. How should I proceed?

A: Plants often wilt following division as they become used to their new surroundings. To lessen stress, make sure the just separated plants receive enough water and are situated in an area with indirect sunshine. The plants should recuperate and start to flourish in their new locations with the right care.

CHAPTER FIVE

CULLING AND APPLYING BEE BALM

Knowing When To Gather

Timely harvesting of bee balm is essential to get the optimal flavor and medicinal benefits from your plants. Gathering bee balm is usually best done when the blooms are completely blossomed but not yet wilting. Depending on where you live and the kind of bee balm you're cultivating, this is typically in the middle to late summer. When your plants are ready to be harvested, keep an eye out for bright, fully-opened blossoms.

When the plant is in full bloom and has reached its most fragrant point, it's also time to harvest bee balm. The lovely aroma of bee balm is well-known, and you may maximize the flavor and fragrance of your harvest by gathering it at the height of scent.

When deciding when to harvest, pay attention to how strong the aroma is and utilize that information to your advantage.

Methods For Gathering Bee Balm

There are a few methods you can employ to guarantee a good harvest of bee balm without endangering the plant.

One way is to carefully clip the stems just above the leaves with pruning shears or scissors. This guarantees that the plant will continue to flourish by enabling you to gather the blossoms without upsetting the rest of the plant.

As an alternative, you can use your fingers to carefully pinch off individual blossoms from the stalk. If you only need a few blooms at a time or have fewer harvests, this strategy works well for you.

The taste and look of the flowers may be impacted by handling them roughly and bruising or injuring them.

Using the same methods as above, concentrate on snipping or pinching the leaves from the stalks if you're gathering bee balm for its leaves rather than its flowers.

The best flavor and texture come from harvesting the leaves when they're still young and fragile, so keep a close eye on your plants and harvest them as soon as they reach maturity.

Medicinal And Culinary Applications Of Bee Balm

Bee balm is a useful addition to any garden because it is a multipurpose herb with both culinary and medicinal benefits. In the culinary arts, bee balm infuses distinct flavors into salads, soups, teas, and sweets. The leaves and blooms offer a delicious flavor to savory and sweet recipes because of their lemony, minty flavor with undertones of oregano and thyme.

Infusing oils or vinegar with bee balm is a common culinary application. To infuse the fresh blossoms or leaves with the flavor and perfume of the herb, simply steep them in your favorite oil or vinegar for a few weeks.

For an additional flavor boost, use the infused oil or vinegar in marinades, salad dressings, or as a last touch on your favorite recipes.

Bee balm is used in cooking and has a long history of therapeutic applications. It can be used to treat a range of illnesses since it includes chemicals with antioxidant, antibacterial, and anti-inflammatory qualities.

Bee balm tea, which is prepared by steeping the leaves or flowers in hot water, is frequently used to ease digestive problems, ease sore throats, and lessen anxiety and tension.

Bee balm can be harvested for its culinary or medicinal uses; however, to maintain its flavor and efficacy, it must be stored correctly.

The most popular technique to preserve leaves and flowers is to dry them in a well-ventilated place out of the sun; alternatively, you can freeze them or turn them into tinctures or extracts for long-term storage. Bee balm has several benefits that you may enjoy all year long if you pick it up at the correct time and store it in your kitchen or medicine cabinet.

CHAPTER SIX

INCLUDING BEES IN THE DESIGN OF LANDSCAPE

Bee balm may be a beautiful addition to any landscape design because of its vivid colors and pleasant scent. Whether you choose a more modern or traditional garden, adding bee balm to your outdoor space can enhance its aesthetic appeal and practicality. Here are some pointers to help you include bee balm into your landscape design seamlessly.

Selecting The Ideal Site:

Take into account your garden's sunlight and soil conditions before growing bee balm. Bee balm grows best in well-drained soil with full to partial shade. To keep your bee balm plants healthy and vibrant, choose a spot with enough sunlight and well-drained soil.

Layering and Grouping: Strategic grouping and positioning are key components of using bee balm in your landscape design. To add visual appeal and impact, try growing bee balm in drifts or clusters. To give your garden beds more depth, you can also stack bee balm with other plants that have different heights and textures.

Combining Bee Balm with Other Perennials:

A variety of perennials, such as sedums, black-eyed Susans, and coneflowers, go well along with bee balm. You can get beautiful color combinations and draw even more pollinators into your garden by combining bee balm with plants that complement one another.

Bee balm can be used to provide well-defined borders and boundaries for your garden beds. Bee balm's tall, erect stems can act as organic dividers, defining distinct areas of your landscape and giving your design structure.

Seasonal Interest: Bee balm's durable flowers are one of the benefits of using them in your landscape design. Bee balm can bloom from early summer to autumn, depending on the variety, and add color and interest all through the growing season.

Maintenance: Make sure to give your bee balm enough water, especially during dry spells, and remove any dead or faded blooms to promote continued flowering if you want to keep it looking its best. Bee balm can also be divided every couple of years to help renew the plants and avoid overpopulation.

Design Suggestions For Using Bee Balm

There are a lot of creative and expressive options when using bee balm in your landscape design. Bee balm may enhance the beauty, color, and ecological value of your landscape, regardless of whether you're creating a formal garden or a more relaxed outdoor area. To help you make the most of this adaptable plant, consider the following design advice:

Color Coordination: Take into account how the hues of the bee balm kinds you choose for your garden will clash or complement those of the other plants in your landscape. For instance, adding different hues of bee balm can give depth and complexity to your color palette, while matching red bee balm with white or purple flowers can produce a dramatic visual effect.

Height and Scale: Bee balm is available in a variety of heights, ranging from little cultivars that are good as edging to large cultivars that serve as stunning focal pieces. When choosing bee balm kinds for your garden design, take into account the overall size of the area and the scale of the garden. Shorter kinds can be used as accents or ground covers, while taller versions can be positioned towards the back of a border or mixed in with other tall perennials.

Repetition and Unity: Adding recurring components to your landscape can help it feel cohesive and cohesive. To create visual continuity and guide the eye across your garden, try planting bee balm in several spots. When paired with other design components like paths, hardscape features, or structural plantings, this repetition may be quite effective.

Seasonal Variation: Bee balm is prized for its fragrant leaf, which emits a pleasing perfume when crushed or stroked, in addition to its vibrant blossoms. When using bee balm in your design, take into account how the seasons will affect the product's look and aroma. All year-long visual appeal and interest can be maintained in your garden by including species that provide complimentary seasonal interest.

Wildlife Habitat: Bees, butterflies, and hummingbirds are among the pollinators who like bee balm. Incorporating bee balm into wildlife-friendly garden designs that offer food, shelter, and habitat for beneficial insects and birds can optimize its ecological benefits.

One way to promote local ecosystems and take in the close-up beauty of nature is to cultivate a landscape that is both diversified and biodiverse.

Functional Design: Bee balm has functional uses in the environment aside from its visual attractiveness. Bee balm can be planted next to fruit trees or vegetable gardens to draw pollinators that boost crop production.

It can also be planted next to outdoor seating areas to give guests a sensory experience thanks to its fragrant foliage and vibrant blossoms.

Ideas For Companion Planting

The gardening practice of companion planting is grouping various plant species to promote growth, ward off pests, and enhance general garden health. There are a lot of companion planting suggestions for bee balm that you should take into account to optimize its advantages and develop a flourishing garden ecology. To help you with your garden design, consider the following companion planting ideas:

Bringing in Pollinators:

With its nectar-rich blossoms, bee balm is well known for drawing in bees, butterflies, and other pollinators. Consider growing bee balm next to other pollinator-friendly plants like lavender, salvia, and coneflowers to increase its efficacy as a pollinator magnet. These companion plants will contribute to the development of a varied and biodiverse garden that is home to a variety of helpful insects.

Keeping Pests Away:

Bee balm not only draws pollinators but also has built-in insect-repelling qualities that can shield nearby plants from pests. Aphids, whiteflies, and spider mites are among the typical garden pests that can be avoided by planting bee balm close to crops that are susceptible to them, such as tomatoes, peppers, and cucumbers. Bee balm is a useful companion plant for pest control because of its inherent deterrent properties, which come from the aromatic oils it releases.

Harmonious Colours:

Think about how the colors of the plants will interact and complement one another in the landscape when choosing companion plants for bee balm. For instance, you can add visual appeal to your landscape by creating a bold and harmonious color scheme with purple bee balm and yellow daisies or orange marigolds. Additionally, combining various textures and colors can help create eye-catching planting arrangements that highlight the unique qualities of each plant.

Seasonal Interest: The seasonal interest of various plants should be taken into account when designing companion planting plans. You can make sure that your garden stays interesting and visually appealing throughout the year by choosing companion plants that bloom at different periods. For instance, you can create a continuous display of color from spring through autumn by combining late-blooming perennials like bee balm with early-blooming bulbs like crocuses and daffodils.

Functional Pairings: Companion planting has several uses beyond just improving the appearance of a garden. For instance, by drawing pollinators and keeping pests away, bee balm planted next to herbs like basil, rosemary, and thyme can enhance their flavor and vigor. Similar to this, bee balm planted close to berry bushes or fruit trees will aid in increasing fruit set and pollination, which will enhance yields when harvest time comes.

Companion Planting Guilds: Establishing planting guilds, or communities of plants that promote each other's growth and health, is a more all-encompassing method of companion planting. In these guilds, bee balm can play a key role, offering the surrounding plants both aesthetic appeal and ecological advantages. Try out several plant combinations to determine which companion planting strategies are most effective in your garden.

Bee Balm: A Tool For Bee-Friendly Garden Design

Not only would a bee-friendly garden help pollinators, but it will also improve the general health and vibrancy of your outdoor area. Bee balm is a great way to draw butterflies, bees, and other beneficial insects to your garden because of its vibrant blossoms and nectar-rich flowers. Here are some pointers for using bee balm to create gardens that are bee-friendly:

Using Plants to Promote Diversity:

Include a wide variety of plants in your garden to give different pollinators food and a place to live. Add blooming herbs, perennials, shrubs, and trees that bloom at various times of the year in addition to bee balm. For bees and other pollinators, this guarantees a consistent supply of nectar and pollen from spring through autumn.

Selecting Local Plants:

Native plants are excellent candidates for bee-friendly gardening since they are well-suited to the local climate and soil conditions.

Choose native plants that attract bees and other pollinators as your top priority when choosing plants for your garden. Certain native varieties of bee balm, like Monarda didyma and Monarda fistulosa, are especially well-suited to sustain pollinator numbers in the area.

Providing Sources of Water:

Especially in the sweltering summer months, bees require access to clean water for drinking and cooling their colonies. To make sure that bees have access to water when they need it, include shallow water sources in your garden, such as birdbaths, saucers filled with pebbles, or tiny ponds. To protect pollinators, make sure the water is pure and uncontaminated.

Steer clear of pesticides:

It is crucial to keep pesticides out of your garden whenever you can because they can affect pollinators like bees. To keep pest populations in check, choose natural pest management techniques like hand-picking pests, applying insecticidal soaps, or introducing beneficial insects like ladybirds and lacewings. You may make your garden a safer and friendlier place for bees by using less pesticides.

Building a Shelter

For them to nest and overwinter, bees need protected places, so make sure your garden has enough habitat. For bee species that nest in cavities, give nesting boxes or bee hotels; for ground-nesting bees, provide sections of bare ground; and for additional cover and protection, incorporate dense shrubs or hedgerows. You may draw a wide variety of bee species to your garden by providing different options for them to build nests.

Teaching Others:

Inform people in your neighborhood about the value of bee-friendly gardening and inspire them to establish pollinator-friendly spaces in their yards.

To increase public knowledge of the risks pollinators face and the actions that individuals can take to help protect them, host workshops, provide resources, and take part in local conservation initiatives. Together, we can build a network of gardens that are beneficial to bees and will sustain healthy pollinator populations for many years to come.

CHAPTER SEVEN

IN COMMERCIAL GARDENING, BEE BALM

Growing Up: Methods Of Commercial Cultivation

It takes meticulous planning and execution to get from a backyard hobby to a commercial enterprise when growing bee balm cultivation. The following are crucial actions to take into account while turning your bee balm garden into a business venture:

Site Selection: Pick a spot with well-drained soil and lots of sunshine. Bee balm can withstand little shade, but it prefers full sun. Make sure it is simple to plant, maintain, and harvest the area.

Soil Preparation: Test the soil to ascertain its pH and nutrient content. Bee balm thrives in well-drained, slightly acidic soil over neutral soil. Adjust the soil as necessary to give the plants the best possible growing environment.

Select the propagation strategy for your bee balm plants. They can be multiplied by divisions, cuttings, or seeds. Select the approach that best meets your needs by taking into account the time and resources needed for each.

Planting:

In the prepared soil, plant bee balm by the spacing guidelines for the particular variety you are cultivating. After planting, give the plants lots of water to help them form roots.

Maintenance: To keep your bee balm plants healthy and vigorous, put in place a regular maintenance regimen. This includes fertilizing, watering, and keeping an eye out for illnesses and pests. As needed, thin the plants to allow for better airflow and avoid crowding.

Gathering: To extract the most essential oils, gather bee balm blossoms at their peak blooming period. Cut off the flower stalks with sharp scissors or pruning shears, leaving some foliage to aid in the plant's growth.

Post-Harvest Handling: Take extra care when handling harvested bee balm blossoms to maintain their freshness and quality. Take out any faded or broken flowers, and keep the rest of the bouquets out of direct sunlight in a cool, dry spot.

Processing and Packaging: Take into account value-added processing choices for your bee balm products, such as extracting essential oils for aromatherapy or drying the flowers for tea.

Stow the goods in eye-catching and useful containers that will both draw in customers and preserve their quality.

These commercial cultivation techniques will help you grow your bee balm gardening business and create marketable, high-quality products.

Promoting Products Made With Bee Balm

To stand out from the competition and reach your target market, marketing bee balm products demands a smart strategy. The following are some successful marketing techniques to think about:

Determine Your Target Market: Recognise the characteristics and inclinations of your intended audience. Take into account variables like age, gender, lifestyle, and shopping habits to adjust your marketing strategies appropriately.

Emphasize the special features of your bee balm goods, such as its all-natural composition, advantageous health effects, or environmentally friendly manufacturing processes. Make sure that your branding, packaging, and advertising materials all highlight these distinctive selling characteristics.

Make Use of Online Platforms: Increase audience reach and boost sales by utilizing the strength of online marketing channels. To promote your bee balm items and ease online purchases, create a polished website or e-commerce store. Engage potential customers and raise brand recognition through influencer partnerships, email marketing, and social media platforms.

Engage in Farmers Markets and Craft Fairs: To establish face-to-face contact with clients, display your bee balm products at neighborhood farmers markets, craft fairs, and community gatherings. To draw interest and increase sales, provide samples, demos, and exclusive offers.

Work Together with Retailers and Distributors: To distribute your bee balm goods through wholesale or consignment agreements, team up with retail establishments, health food stores, spas, and gift boutiques.

Establish connections with retailers and give them the tools and resources they need to effectively promote your goods, including marketing materials.

Inform Customers: Inform customers on the advantages of bee balm and the best ways to include it in their everyday routines.

Post educational material on your site, via articles, videos, and social media to establish oneself as a reliable authority in the field.

Get Client comments: Learn about your consumers' needs, preferences, and degree of satisfaction with your bee balm goods by getting their comments. Make adjustments, create new product offerings, and improve your marketing tactics with the help of this feedback.

You may effectively promote your bee balm items and draw devoted clients to your company by putting these marketing ideas into practice.

Regulatory And Legal Aspects

Growing and selling bee balm products requires careful consideration of the legal and regulatory environment. The following are some important things to remember:

Licenses and Permits: Find out which licenses and permits are necessary in your area to run a commercial bee balm garden and market bee balm goods. Depending on where you live and what kind of business you run, these could include business licenses, agricultural permissions, and permits for handling food.

Labeling Requirements: Learn about the laws governing the labeling of herbal items, such as teas made from bee balm, extracts, and essential oils. Make sure that all of the required labeling information—such as ingredient lists, net weight or volume, and contact details—is included on the labels of your products.

Observe quality and safety guidelines to make sure your bee balm goods live up to consumer expectations and legal requirements. Throughout the production process, adhere to good manufacturing practices (GMP) and good agricultural practices (GAP) to uphold strict quality, hygienic, and traceability standards.

Application of Pesticides: Apply pesticides and other agricultural protection agents sensibly and in compliance with regional laws.

 To prove that you comply with regulations, keep thorough records of all pesticide applications, including the kind of pesticide used, application rates, and dates of application.

Environmental Regulations: Take into account how your bee balm cultivation methods will affect the environment and take action to reduce pollution, preserve natural resources, and safeguard biodiversity. To maintain sustainable farming methods, abide by environmental standards governing waste management, habitat conservation, and water consumption.

Intellectual Property Rights: To protect your bee balm variations, branding, and proprietary formulations, protect your intellectual property rights, including patents, trademarks, and copyrights. To learn more about your rights and obligations about the enforcement and protection of intellectual property, speak with legal professionals.

You can minimize potential risks to your bee balm gardening business and ensure compliance with applicable laws and regulations by proactively addressing these legal and regulatory factors.

CHAPTER EIGHT

COMMONLY ASKED QUESTIONS OR FAQS

What is Bee Balm?

Native to North America, bee balm—scientifically known as Monarda—is a stunning floral plant. It is loved for its vivid flowers and fragrant leaves and is a member of the mint family. The plant usually grows in clusters, and its vibrant tubular flowers draw pollinators such as hummingbirds, butterflies, and bees. Bee balm adds a beautiful flash of color to any landscape because it comes in a variety of tints, including red, pink, purple, and white.

How Is Bee Balm Planted?

Bee balm planting is an easy task that takes very little time. First, choose a spot that receives full to partial sunlight, well-draining soil, and both. Make sure the bee balm is at the same depth as it was in its container by excavating a hole that is just a little bit bigger than the root ball of the plant. To help the plant form roots, properly water it and pat the dirt around it gently. Multiple bee balm plants should be spaced 18 to 24 inches apart to allow for proper growth and airflow.

What Handling Is Needed for Bee Balm?

Because bee balm requires little upkeep, it's a great option for novice and experienced gardeners alike. Make sure the soil stays regularly damp but not soggy, especially in the heat. Mulching the area surrounding the plant's base can aid in moisture retention and inhibit the growth of weeds. Regularly deadheading spent flowers promotes ongoing blooming and inhibits self-seeding, however, some gardeners might be happy to see bee balm proliferate in their space. Furthermore, splitting established clumps every few years helps revitalize the plant and encourage more robust development.

How Can Bee Balm Be Propagated?

Using seed sowing or division, bee balm can be propagated easily. Dig out the root ball of an existing plant and carefully split it into smaller portions, making sure that each division has numerous healthy shoots and roots attached.

To aid in the establishment, replant the divisions in the prepared soil and give them plenty of water. As an alternative, gather seeds from fully grown bee balm blossoms and plant them straight into the garden in the autumn. Until seedlings appear and the seeds germinate, keep the soil continually moist.

Common Questions Concerning Bee Balm

What Advantages Does Bee Balm Growing Offer?

Growing bee balm has many advantages, chief among them being its capacity to draw pollinators into the garden. The plant's vibrant blooms attract bees, butterflies, and hummingbirds, making it an invaluable addition to any landscape that supports pollinators. Additionally, because of its antibacterial and anti-inflammatory qualities, bee balm is frequently employed in herbal therapy. The leaves can be applied topically to soothe minor skin irritations or brewed into a calming tea.

Does Bee Balm Suit Diseases or Pests?

Although most pests and diseases may be avoided by using bee balm, it can occasionally face problems like powdery mildew or spider mites.

Make sure to provide enough space between plants to allow for good circulation and steer clear of overhead watering, which can encourage the spread of fungi, to prevent these issues.

Regularly check the foliage for symptoms of disease or pests, and take quick action to remedy any problems as soon as necessary with chemical or organic treatments.

Can Bee Balm Be Grown in Containers?

Bee balm is a great option for gardeners with limited space or those wishing to add a dash of color to their patios, balconies, or decks because it can be grown effectively in pots. Pick out a sizable container with plenty of drainage holes, then add a potting mix that drains well to it. To maintain the soil uniformly moist, set the container in a sunny spot and give it regular irrigations. Keep in mind that fertilizing on occasion and deadheading spent blooms can encourage healthy growth and bloom.

Troubleshooting Manual For Typical Problems

Issue: Turning Yellow Leaves

Bee balm plants with yellowing leaves may be experiencing several problems, such as pest infestation, nutrient deficits, or overwatering. To keep the soil constantly moist but not soggy, check the moisture content of the soil and modify irrigation as necessary. Repotting the plant into a well-draining mix or enhancing drainage are two options to consider if the soil is too damp. To make up for any nutrient deficits, fertilize the plant additionally using a balanced, water-soluble fertilizer. Examine the leaves for indications of pests like spider mites or aphids, and use neem oil or insecticidal soap as necessary.

Issue: granular mildew

Bee balm plants are susceptible to powdery mildew, a common fungal disease that thrives in damp or poorly ventilated environments. Make sure plants are properly spaced apart to allow for ventilation and steer clear of overhead watering, which can encourage the growth of fungi, to prevent powdery mildew. Should powdery mildew arise, remove any impacted foliage and use a fungicidal spray specifically designed to manage powdery mildew on the plant. To further guard against future outbreaks, think about using a preventive fungicide early in the growth season.

Issue: Sagging or Tilting

Bee balm plants that have wilted or drooping foliage may be signs of root rot brought on by overwatering. To keep the soil constantly moist but not soggy, check the moisture content of the soil and modify irrigation as necessary.

Reduce the frequency of watering and enhance drainage by amending the soil with sand or perlite if the soil is too damp.

To encourage new growth, trim any withered or damaged leaves. You could also think about adding an organic mulch layer to help control temperature and soil moisture retention.

Professional Hints And Counsel

Motivate pollinators

Plant bee balm in huge drifts or clusters rather than single specimens to optimize its potential to draw pollinators. This produces an eye-catching show that hummingbirds, butterflies, and bees can't resist. Include more plants in your garden design that are good for pollinators to provide a steady supply of nectar and pollen all through the growing season.

Regular Pruning

Frequent pruning lowers the danger of disease and improves ventilation while also assisting in maintaining the form and aesthetics of bee balm plants.

Eliminate wasted flowers as soon as possible to promote ongoing flowering and stop self-seeding. To revitalize the plant and encourage new growth from the base, clip down any lanky or crowded growth as well.

Collect Leaves for Tea

Throughout the growing season, bee balm leaves can be collected and dried for use in herbal tea mixtures. Just cut the leaves off as needed, being careful not to take more than one-third of the foliage at once from the plant.

After giving the leaves a thorough rinse, let them air dry in a place with good ventilation and shade from the sun. After the leaves are dry, put them in an airtight container so you can use them later to make calming, fragrant teas.

CHAPTER NINE

COMMON QUESTIONS AND ASSISTANCE

Overcoming Typical Obstacles In Bee Balm Gardening

Newcomers to the realm of bee balm gardening frequently run into several obstacles that might leave them perplexed. Soil quality is one prevalent problem. Bee balm grows best in well-draining soil that ranges in pH from slightly acidic to neutral. Compacted, heavy soil can prevent appropriate drainage and cause root rot. To remedy this, think about enhancing the texture and drainage of your soil by enriching it with organic materials like compost or peat moss.

Overwatering is another issue that gardeners deal with. Bee balm does not like wet circumstances, although it does like normal moisture. Soil that is too damp might suffocate roots and result in fungal infections. Water your bee balm plants deeply but sparingly, letting the top inch of soil dry out in between applications, to avoid this. By minimizing the amount of moisture on the leaves and directing water directly to the roots, soaker hoses, and drip irrigation systems can help lower the risk of disease development.

Bee balm planting can sometimes be challenging because of pests. Powdery mildew, spider mites, and aphids are a few typical pests that can harm your plants. Think about using integrated pest management strategies to get rid of these pests, like hand-picking insects, applying an insecticidal soap spray, or putting beneficial insects like lacewings and ladybirds in your garden. Additionally, fungus infections like powdery mildew can be avoided by keeping proper airflow around your plants by spacing them apart and preventing overcrowding.

Lastly, some gardeners might find it difficult to get strong growth and colorful blossoms from their bee balm plants. A common explanation for this is a lack of sunlight. For best growth and flowering, bee balm needs at least six hours of direct sunlight per day, however, it may grow in both full and partial shade. If your plants are not growing as well as you would like, assess how much sunshine they receive and, if required, think about moving them to a more sunny spot.

Through the proactive resolution of these prevalent issues, you can surmount obstacles in bee balm gardening and foster robust, flourishing flora in your landscape.

Techniques For Achievement

Growing bee balm successfully calls for careful planning, nurturing, and close attention to detail. The correct cultivar choice for your garden is a crucial component of any successful approach. Selecting the right kind for your climate, soil type, and aesthetic tastes is crucial because there are many to select from, each with its distinctive colors, sizes, and growth patterns. Doing your homework on various cultivars and their unique needs will help you make an informed choice and position yourself for success right away.

Maintaining sufficient space between bee balm plants is another tactic. Dense populations can impede ventilation and encourage the spread of disease, which can result in problems like powdery mildew. area the individual bee balm plants at least 18 to 24 inches apart when planting them to ensure adequate air circulation and growth area. This lessens the chance of illness and lessens the competition between nearby plants for sunlight and nutrients.

Success in bee balm gardening also depends on regular upkeep. This entails doing things like pulling weeds, deadheading spent blooms, and keeping an eye out for pests and illnesses. While weed eradication reduces the chance of harboring pests and illnesses and prevents competition for resources, deadheading promotes continuous blooming throughout the season. You can keep your bee balm plants healthy and vibrant by being watchful and taking quick action when necessary.

Additionally, to improve the aesthetics of your bee balm garden and draw helpful insects, think about adding companion plants. Coneflowers, salvias, and lavender are examples of flowers that not only visually enhance bee balm but also give pollinators access to other sources of nectar and pollen. Herbs like basil, thyme, and oregano can also draw beneficial insects and discourage pests, fostering a more resilient and well-balanced ecosystem in your garden.

Last but not least, mulch the area surrounding your bee balm plants to retain moisture, keep weeds at bay, and control soil temperature. These advantages can be obtained over time by using organic mulches, such as compost, straw, or shredded bark, which help improve soil fertility and structure. Don't pile mulch up against the stems of your plants, since this might encourage disease and rot. Instead, apply a layer of mulch around the base of your plants.

You can optimize your bee balm garden's aesthetic appeal and yield by putting these success strategies into practice. You'll also be able to enjoy colorful blossoms and robust plants all during the growing season.

Resources For Additional Education And Assistance

There are several resources out there to help and guide bee balm gardeners who want to expand their knowledge and proficiency in the art. Books and magazines on gardening that are devoted to perennial flowers, or more especially, bee balm, are a great resource. These books frequently include thorough instructions on growing bee balm, covering planting, upkeep, propagation, and common problem-solving. Seek out books written by respectable authors or institutions with experience in perennial gardening to make sure the material you're reading is factual and trustworthy.

For information and experience sharing with other bee balm fans, online forums, and gardening communities can also be very helpful. Active communities can be found on websites such as GardenWeb, Houzz, or the gardening subreddit on Reddit. These forums allow you to interact with like-minded gardeners and exchange questions and tips on bee balm and perennial gardening in general. Taking part in these forums can offer insightful opinions, helpful advice, and encouragement from seasoned gardeners who have encountered comparable difficulties and achievements in their gardens.

For practical courses, lessons, or seminars on bee balm planting, think about contacting your local extension office, botanical gardens, or gardening clubs. Expert speakers who can offer helpful guidance, demonstrations, and Q&A sessions customized to the climate, soil conditions, and pest concerns of your region are frequently featured in these educational events. Making connections with knowledgeable locals and other gardeners can help you learn more and build a sense of support and community among like-minded people.

Finally, don't ignore internet resources like the websites of horticultural associations, botanical gardens, and university extension programs. These resources frequently offer fact sheets, articles, videos, and research-based information on the production and maintenance of bee balm. In particular, university extension websites include state-specific materials and science-based advice to help you overcome obstacles in your bee balm garden and make well-informed decisions.

You can continue to develop and enhance your bee balm gardening abilities and ultimately achieve more success and fulfillment in your gardening pursuits by making use of these resources for additional education and assistance.

www.ingramcontent.com/pod-product-compliance
Lightning Source LLC
Chambersburg PA
CBHW070313230526
45470CB00002B/850